A Pz.Kpfw.I Ausf.A (3.Series/La.S) converted to Panzer Schulfahrzeuge and photographed in June 1940. It is unusual in having been fitted with a radio indicating that it was probably a repair or munitions vehicle. Note that the chap on the right is listening on his headphones.

L.Archer

A Schulfahrzeuge (driver training vehicle) with Fgst.Nr.14650 from the 5b./La.S. series, completed by Krupp-Grusonwerk by the Spring of 1937, was still in service with a maintenance unit belonging to the 20.Panzer-Division on the eastern front in 1943. While the handrails on the sides and bent around to the front are original, it has been 'upgraded' by adding a padded bench seat, radio set, and a frame for fitting a canvas cover for protection against the elements. The 'Notek' blackout light is mounted much higher than its normal position on a track guard. **3x L.Archer**

The Pz.Kpfw.I Ausf.A could climb a gradient of up to 30 degrees. The articulation of the suspension units at this extreme angle is interesting as the front roadwheel looks to be nearing the end of its travel. The little tank would have made quite a noise too, as the muffler is no longer connected to the exhaust and staining is visible around the hole in the side of the engine deck. As noted on the back of the original photograph and verified by the shape of the track leaving the drive sprocket, this Pz.Kpfw.I is climbing this steep slope - not descending.

L.Archer

4

What a rarity. The camouflage pattern with all three colours: RAL 17 (Erdgelb), RAL 18 (Braun) and RAL 28 (Grün) of the Buntfarbenanstrich can be seen on this Pz.Kpfw.I Ausf.A from the 3./La.S. series. Usually 35mm photos don't have sufficient contrast to show all three. As instructed in the manual, to avoid creating a repeating pattern when in line or in a column, no two artillery pieces, trucks, tanks, or other field equipment, were to be painted with the same camouflage pattern.

2x L.Archer

Pz.Kpfw.I hopscotch. A Pz.Kpfw.I Ausf.A (from the 3./La.S. series) crosses over a ravine on a wooden bridging section mounted on another Pz.Kpfw.I (which has the turret removed). The white Balkenkreuz on the travelling Pz.Kpfw.I are freshly applied, these were used between July and October 1939, which would date this photo from that period, the fact that all of the ports and hatches are open on the tank would support this. A chassis number is not visible on the vehicle, although an unreadable one can be seen on the roadwheel jack.
L.Archer

This blurred example of a Pz.Kpfw.I Ausf.A on campaign has been 'personalised' by the crew with the addition of two rows of spare tracks across the front and a backfitted 'Notek' blackout light mounted on the nose rather than the track guard. The roadwheel jack has been moved from the fender to the glacis plate with a stowage box replacing it. A smoke grenade launcher has been fitted onto the left rear track guard. This was a feature of Pz.Rgt.21 in the 20.Panzer-Division, which due to a shortage of Pz.Kpfw.II had 44 Pz.Kpfw.I at the start of Operation Barbarossa in June 1941.

L.Archer

A turretless Pz.Kpfw.I Ausf.A in use as a munitions carrier or maintenance/repair vehicle. The back of the photo is humourously captioned "unser Sturmpanzer" and was most likely taken during an exercise. While a crewmember in the fighting compartment (or is he behind the vehicle?) throws what looks like a sand bag, the driver peers out of his visor. A careful look at the position of the drive sprocket teeth in the tracks shows the vehicle to have turned into this position.

L.Archer

Waiting for the return of a repaired or replacement drive sprocket for this Pz.Kpfw.I Ausf.A, two troopers take an opportunity to get in 40 winks - even though they were probably left behind to guard against pilfering by opportunists, like these disappointed infantry. The snake of tracks in front of the tank is interesting as it has a mixture of open and solid guide horns. Part of a tactical sign is visible on the driver's front plate, just above the foliage.

L.Archer

9

This Pz.Kpfw.I Ausf.A (Fgst.Nr.9301 from the 2./La.S. chassis series assembled by Krupp-Grusonwerk) has a narrow brake access hatch on the glacis plate. One of the roadwheels is missing and the tank has taken on a sagged look. The plain white Balkenkreuze, in this instance purposely dirtied, indicate that the photograph was taken during the Polish campaign. A hand waves from the opened driver's visor.

L.Archer

This Pz.Kpfw.I Ausf.B ran over a mine in Warsaw, September 1939 which blew away the first roadwheel and made a mess of the tracks. Recovery of the stricken tank has been attempted with a towrope in place on the rear towing pintle and both the jack block and tankers bar are lying at the rear of the vehicle. A smoke grenade rack is fitted at the rear of the tank, behind the exhaust muffler.

L.Archer

Armour designed to prevent penetration by armour-piercing bullets fired by rifles or machineguns could not (and was not intended to) withstand a direct hit from an artillery shell that has obliterated the superstructure front and damaged the turret of this Pz.Kpfw.I Ausf.B. German soldiers somberly look over the wreck, probably pitying those who fought in the vehicle. Through the hole in the front the driver's seat is visible, as are some of the details under the turret race. The object on top of the turret is one of the M.G.13k machineguns.

L.Archer

A look at the side of the wreck reveals more damage to the fighting compartment and a large swathe of missing turret. One of the M.G.13k machineguns dangles from the turret side, held only by its firing controls. What looks like an anti-tank penetration in the side of the fighting compartment is in fact where a section of armour plate has been blown out; the missing triangular section is leaning against the engine compartment with a lifting hook at the top.
L.Archer

A 1.Serie/kl.Pz.Bef.Wg (Sd.Kfz.265) drives past a motorised column. As was the custom when it was assembled in 1936, it has a three-tone Buntfarbenanstrich camouflage scheme. However, unlike the Pz.Kpfw.I Ausf.A with distinct edges at the borders, the sprayed on paint was allowed to spatter so that the colours melded together. Due to delays in producing the machinegun ball mounts, many of the 1st series kl.Pz.Bef.Wg. were delivered without them. These 1st series of kl.Pz.Bef.Wg. are also unique in having a twin-hatch for the commander instead of a cupola.

L.Archer

As shown here, all of the second and third series of kl.Pz.Bef.Wg. were not made the same. The first one with tactical number '401' still has a reinforcing strip of armour bolted to the superstructure side directly above the track guard. The second one does not have this reinforcement, but instead features a pistol port in the right rear slanted superstructure plate. It appears that these four are out on a training exercise (perhaps in the flat terrain of the Luneberger Heide near Munster?) with a higher-ranking officer officiating. **2x W.Auerbach**

One of the kl.Pz.Bef.Wg (Sd.Kfz.265) has paused on a rise to have its photo taken. Large-scale manoeuvres, vital to combined arms training, were a luxury seldom repeated once the war started. Nonetheless, the lessons learned during these early field exercises contributed greatly to the operational excellence of the German Panzertruppen in the early campaigns. Note the unusual placement of the 'Notek' blackout light on the front of the superstructure.   **W.Auerbach**

An 8.Panzer-Division Pz.Kpfw.II Ausf.A or B, complete with a Russian prisoner, takes a break during the Russian campaign. It has been backfitted with 20mm armour on the hull front, superstructure front, and turret front during 1940 and a commander's cupola with 8 periscopes a year later. The two stowage boxes on the fender were backfitted as well. One of the boxes carries the tactical number '42'. Note the marks left on the roadwheel tyres by the tracks and the armour patch welded to the lower nose plate.

L.Archer

The photographer of this Pz.Kpfw.II Ausf.A was lucky enough to catch the light just right so we can see the factory applied RAL 46 (Dunkelgrau) and RAL 45 (Dunkelbraun) colour scheme. The guide horns of the tracks have worn away the aluminium rims on the roadwheels. Behind is a kl.Pz.Bef.Wg (Sd.Kfz.265).

G. Westbrook

1940. A crewman sits atop a mint condition Pz.Kpfw.II Ausf.b. It carries a chassis number of 21071 in black on the nose and the tactical sign of 19.Panzer-Division on the driver's front plate. The crewman's footprints are visible on the glacis plate. **L.Archer**

The order to change the Balkenkreuz specified that they only be stencilled on the sides and rear of the superstructure. Therefore, this Pz.Kpfw.II Ausf.b belongs to a Pz.Ers.u.Ausb.Abt. (replacement and training battalion). As a Panzer used for training, it is amazing that it still has both the front and rear flaps on the track guard. The AA mount was set at an angle on purpose to swing over other objects on the vehicle. The circular object in front of the AA MG mount on the superstructure side is the end cap of the fire extinguisher. **2x L.Archer**

A Pz.Kpfw.II Ausf.a, or Ausf.b with an Ausf.a drive sprocket, has come to grief in a French town. It carries a large white rectangle for aircraft recognition on the engine deck, a Balkenkreuz on the rear superstructure and a smoke grenade rack behind the exhaust muffler. Here the early suspension was obviously no match for a French kerb. **W.Auerbach**

A Pz.Kpfw.II Ausf.B or C has toppled from its heavy-duty truck. Only Pz.Abt.66 and 67 with the 2. and 3.leichte-Division remained 'transported' during the Polish campaign. Since tracked vehicles could not keep up with wheeled vehicles on the road, this was an attempt to get the Panzers forward as quickly as possible to clear out resistance that held up the unarmoured motorised infantry. Note the massive chocks installed in the truck bed.

W.Auerbach

A Landser poses in front of a wrecked Pz.Kpfw.II Ausf.A or B, backfitted with extra armour on the hull front, superstructure front and turret front. The engine deck has gone, as has the radio antenna, along with its trough - which should be on the left track guard. The middle roadwheels and return rollers have been burnt and are missing much of their tyres. The fact that there is no ash or charring on the ground leads us to conclude that it was towed to this location. In the background is another Pz.Kpfw.II, upside down and stripped of its roadwheels.
L.Archer

German soldiers look over the hulk of a knocked out Pz.Kpfw.II Ausf.c near Saloniki, Greece. It is in a sorry state with the tracks missing, damage to the left drive sprocket and a penetration in the front of the turret visible in this view. An internal explosion has blown out the driver's vision ports and fire has removed paint from the turret. In 1940 the tank was modified by adding the extra armour (on the hull front, superstructure front, and turret front) and in 1941 modified by adding the cupola with periscopes.

L.Archer

25

RPi°

Left: On the dust-cover of the coaxial M.G.34 is the chassis number 20007, making this a Pz.Kpfw.II Ausf.a/1. The large 'RPi' tactical number stands for Pioneer-Zug in the Regimental headquarters. In front of this a war wound has been patched up with a plug of armour plate. The two small hinged ports on the superstructure side are fuel filler ports.　　　**L.Archer**

Above: A Pz.Kpfw.II Ausf.b of Pz.Rgt.7, 10.Panzer-Division has turned over, but how? And where? Note the distinctive Bison insignia stencilled onto the turret side and the opened signal port in the commander's hatch. The spare roadwheel on the fender has been wired on preventing it from falling off during travel - or while rolling over.　　　**L.Archer**

Two shots of a Pz.Kpfw.38(t) Ausf.E or F loading ammunition. In the left picture, with the front steering brake hatch open, we can see that it is painted in RAL 7009 (Hellgrau). The crewman standing in the background is checking the belts of 7.92mm M.G.37(t) ammunition. A total of 2700 rounds of 7.92mm and 90 rounds of 3.7cm ammunition could be carried; two opened magazines of 3.7cm are visible in the right photo.

2x L.Archer

A Pz.Bef.Wg.38(t) drives past a cheering crowd in a Baltic state in 1941. As a command tank, the radio operator's ball mounted M.G.37(t) was removed and plated over and extra radio equipment installed - hence the extra antenna. The 3.7cm Kw.K was also removed and replaced with a fixed dummy gun which has an external 'mantlet' clearly seen here. **L.Archer**

A Pz.Kpfw.38(t) Ausf.B with the weapons dismounted and in good condition. It has a 'Jerry can' rack on the engine compartment, which is an indication that it belonged to Pz.Rgt.25 of the famous 7.Panzer-Division commanded by General Rommel. A white rectangle was painted on the rear deck for aircraft recognition. The wheel rims are shiny bare metal from having contact with the inside of the track guide horns. Likewise, the outsides of the guide horns have a distinctive wear pattern on them from contact with the drive sprocket and idler wheel.

2x L.Archer

A Pz.Kpfw.38(t) of Pz.Rgt.25 in the 7.Panzer-Division lies knocked out after crossing the bridge at Landrecies, France on 17 May 1940. It had been hit 15 times by 25mm cannon fire from a Panhard 178 armoured car of 6eme Cuirassiers, 1ère D.L.M, killing the driver and forcing the other three members of the crew to bail out. The Somua S-35 was from 1ère D.L.M; either 4eme Cuirassiers or 4ème Dragons. Quite whose baby is in the pram is anyones guess.

1x W.Auerbach, 1x L.Archer

German troops in discussion in front of a Pz.Kpfw.38(t) Ausf.D belonging to Pz.Rgt.21 of 20.Panzer-Division. As indicated by the small letter 'W' next to the rhomboidal tactical sign on the front and side, this Sd.Anh.115 trailer belongs to the Werkstatt Kompanie.

W.Auerbach

A Pz.Kpfw.38(t) Ausf.G on campaign. It does not carry a national insignia, just the tactical number '214' on a stowage bin attached to the fender. Other stowage bins adorn the fender - all held in place with what looks like string. A 'Jerry can' sits in a rack on the turret side, another on the engine deck. A helmet and a mess tin hang from the turret side.

L.Archer

A column of Pz.Kpfw.38(t), the lead tank an Ausf.B, are about to make a turn, as indicated by the radio operator with his arm out. The steering brake hatch on the glacis has been opened to give a little respite from the heat inside the vehicle. The only visible markings are a 'G' marking on the driver's front plate and a glimpse of a Balkenkreuz on the side of the superstructure. It is interesting to note that rain guards have been fitted over the commander's forward vision block in his cupola and the gunsight.

L.Archer

The Pz.Kpfw.38(t) Ausf.E assigned to the platoon leader of the 4.Zug in 4./Kp/II.Abt/Pz.Rgt.27, 19.Panzer-Division has turned turtle into a ditch. The stowage bin on the track guard either side of the fighting compartment, is a feature of vehicles of this unit, as are the 'Jerry can' racks on the rear track guards and the two spare track links on the front.

L.Archer

A great find. A brand new Pz.Kpfw.38(t) Ausf.A with the two tone camouflage scheme of RAL 46 (Dunkelgrau) with blotches of RAL 45 (Dunkelbraun) clearly visible on the turret side. It is quite unique to have a photo of a Pz.Kpfw.38(t) still with the original factory layout before the units 'improved' the stowage. Originally designed for a crew of three, there wasn't much space for internal stowage when manned by a crew of four. Note the 'plug' to keep the gun clean and the blank adaptor for training fitted to both machineguns.

L. Archer

A gr.Pz.Bef.Wg Ausf.D1 (Sd.Kfz.267 or 268), tactical number 'R02' next to a kl.Pz.Bef.Wg. (Sd.Kfz.265). In this clear view we can see the MP port that replaced the radio operator's ball mounted MG and the MP port next to the drivers side visor. The main armament was replaced by a dummy gun with a dummy machinegun next to it; the sole armament for the turret was a M.G.34 in a ball mount. Note the interlocking upper and lower hull front plates.

G.Westbrook

This gr.Pz.Bef.Wg Ausf.E (Sd.Kfz.266, 267 or 268) makes for an interesting comparison with the example on page 38 and the differences can be appreciated here. The radio operator's MP port is in the open position revealing its lozenge shape and the frame antenna over the engine deck has been removed.

W.Auerbach

A gr.Pz.Bef.Wg Ausf.E (Sd.Kfz.266, 267 or 268), tactical number '1101' assigned to the commander of II.Abt./Pz.Rgt 5 or 6 of the 3.Panzer-Division has run off the road and become well and truly stuck. A large unditching beam has been wired to the track guard, but it will not help much here. Some interesting details are visible, such as the large number of 'Jerry cans' held by wire to the rear of the turret. As the turret of the Pz.Bef.Wg did not rotate, they would not get in the way like they would on a gun tank. Behind the unditching beam we can see the vision and MP ports, while on the turret roof is the circular hatch for the 'Kurbelmast' (telescoping mast for a raised antenna).

L.Archer

A careful look at the back plate reveals the unit insignia for 3.Panzer-Division from 1940 (a rotated 'E') below the Balkenkreuz. The original frame antenna has been removed; just the mounts remain on the engine deck and on the hull rear plate. Note how leaves have collected on the wire screen over the air intakes either side of the engine deck.

L.Archer

Two Landsers check out a gr.Pz.Bef.Wg Ausf.D1 (Sd.Kfz.267 or 268) that has run over a mine in France, leaving its suspension wrecked. This vehicle has been fitted with a smoke grenade rack at the rear of the hull (all of its smoke grenades have been dropped) and it carries a small fascine (bundle of sticks) for crossing ditches and trenches. The small circular hatch for the 'Kurbelmast' can be seen open in the front view. **2x W.Auerbach**

A photo probably taken during the lull between the fall of France and the invasion of Russia, when the number of Panzer Divisions was doubled. Here infantry pose with a Pz.Kpfw.III Ausf.G. This example mounts a 5cm Kw.K first introduced in July 1940, but it retains the older commander's cupola with five vision slits. The crew wear a mix of black Panzer berets and field caps.

W.Auerbach

One of only 15 Pz.Kpfw.III Ausf.C assembled by Daimler-Benz, this one knocked in Poland. It has at least five penetrations from a 3.7cm anti tank gun, although whether these occurred in combat is debatable as three are very tightly grouped. In this front view, we can see that the twin coaxial M.G.34s have been removed, along with that of the radio operator. The makeshift graves for two of the crew on page 45 tell a more sombre story.

2x W.Auerbach

Already stripped of suspension components, a badly battered Pz.Kpfw.III Ausf. F with 5./Z.W. Turm seems to await a coup de grâce from a signal pistol. Details that aid in identification are the cowls for the brake cooling inlets on the glacis plate and splash-guard for the turret ring. Note that the 3.7cm Kw.K appears to have taken a hit and the trough for the radio aerial is absent. The Balkenkreuz on the hull side would indicate it met its fate either in France or later.

L.Archer

How to hold up the convoy - tank style. A Pz.Kpfw.III Ausf.H has thrown its track and can no longer continue (see how the guide horns are on the outside of the idler and roadwheels) stalling Pz.Rgt.31 of 5.Panzer-Division on a narrow mountain pass in Serbia, April 1941. The tank carries spare track in racks on each of the track guards and across the top of the turret.

The Pz.Kpfw.IV Ausf.E following behind has the frame for mounting extra armour on the superstructure front plate in front of the radio operator, but due to delays in producing the extra armour plates some Ausf.E were delivered without them.

L.Archer

Soldiers pose with wrecks at a collection point. On the left is a Pz.Kpfw.III Ausf.G fitted with a 6b./Z.W.Turm armed with a 5cm Kw.K and external mantlet (which is missing the armour guard for the coaxial M.G.34). It has the early style commander's cupola, which has only part of its hatches left. Two hooks have been welded to the driver's front plate to hold spare tracks. The example covered in tracks on the right is also an Ausf.G, but fitted with a 6a./Z.W.Turm armed with a 3.7cm Kw.K, dual coaxial M.G.34s and later style commander's cupola. **W.Auerbach**

A German soldier peers into the escape hatch of a knocked out Pz.Kpfw.III Ausf.J at a railway station 'somewhere in Russia' amid the wreckage of Russian trucks and burnt out buildings. The tank has a camouflage scheme of RAL 7028 (Dunkelgelb), RAL 8017 (Rotbraun), and RAL 6003 (Olivgrün) that replaced the monotone RAL 7021 (Dunkelgrau) in early 1943. Now, whose boots are on the fender of the Russian truck?

L.Archer

Two views of the same Pz.Kpfw.III Ausf.J in service in Russia with Pz.Rg.31 of the 5.Panzer-Division. The photo on this page was the first to be taken and it is interesting to see how the whitewash weathered and in what areas. The devil's head unit identification symbol stencilled on the turret side has not been covered with whitewash showing the RAL 7021 (Dunkelgrau). The tank was whitewashed with the turret front visors shut and now with one open we see the original paint exposed.

2x L.Archer

A rare sight indeed; not one but two Pz.Kpfw.IV fitted with 'Vorpanzer' (spaced armour) on the turret and superstructure front as well as the original 'Zusatzpanzer' (extra armour) bolted onto the superstructure and hull front and sides. The example in the foreground is an Ausf.D. The 20mm thick 'Vorpanzer' on the turret was made of three sections; two 'cheeks' which protected either side of the turret front, with openings for the viewing ports and gunsight, and a curved plate that was bolted to the gun mantlet. The photos on pages 52-55 were taken at Cottbus in 1942 by a soldier training with 6.(Panz).Inf.Ers.Rgt. 'Großdeutschland.'

L.Archer

This is the tank seen in the background of page 52. It is an Ausf.E also with 'Vorpanzer' (spaced armour) on the turret and superstructure front as well as having the original extra armour bolted onto the superstructure front and sides and hull sides. It has a different style of Balkenkreuz and a field fitted stowage bin in a rack on the rear of the turret. Above the roadwheels are three racks presumably to hold spare track links. This photograph graphically shows how the deflector mounted on the gun sleeve folded the radio antenna down.

L.Archer

Above: A Pz.Kpfw.IV Ausf.D with 'Vorpanzer' and a Pz.Kpfw.II backfitted with a commander's cupola, extra armour, and a large stowage bin on the track guards have been pulled out of the maintenance bays for the Werkstatt-Kompanie. **Inset:** A wider view of the scene shows more tanks including a Pz.Kpfw.III.

Right: A field exercise with a smoke screen engulfing the side of a Pz.Kpfw.IV. This is the tank shown on page 53 with the makeshift stowage bin on the turret rear.

L.Archer

2x L.Archer

Above: The vehicles have moved to the Truppenübungsplatz Wandern (today Wędrzyn in Poland), some 140km from Cottbus for firing practice. A crew is shown here cleaning the 7.5cm Kw.K.

L.Archer

Right: The two Pz.Kpfw.IV in position at a firing range at Wandern with a sizeable quantity of crated ammunition for gunnery practice. In the lower photo, an officer is using his binoculars to spot where each round hits. The crew of the left tank have opened the ventilation hatch in front of the commander's cupola.

2x L.Archer

On the left, the Kompanie Chef; Oblt. Ringer observes the gunnery practice; while in the top photo the same tank is shown after firing, the gun having recoiled. In the bottom photo an unknown officer sits on the turret stowage bin as a crew takes its turn in gunnery practice.

3x L.Archer

The gunner has his forehead firmly seated in the headrest of the telescopic T.Z.F.5b gunsight in order to maintain a steady sight picture. Details on the inside of the door such as the thick glass block for the vision slit and the slide for the pistol port are clearly seen in this photo.

L.Archer

Left: A Pz.Kpfw.IV Ausf.C with the internal gun mantlet and pistol port in the superstructure front plate loaded on a railcar at Schermeisel, today Trzemeszno, East of Wędrzyn (Wandern). A Pz.Kpfw.IV Ausf.D uparmoured with 'Vorpanzer' is on the railcar behind. In the foreground is StFw Bauer.   **2x L.Archer**

Right: This Pz.Kpfw.IV Ausf.D from 7.Panzer-Division has been repaired with a flat plate replacing the original fitted with a machinegun ball mount. Part of the tactical number is visible on the opened turret side hatch.   **W.Auerbach**

61

Above: Pz.Kpfw IV Ausf.Bs and a Pz.Kpfw.III Ausf.D of 5.Panzer-Division undergoing repair in the market square at Opatów, Poland in 1939. A soldier leans over a toolbox carrying the unreadable chassis number of the vehicle it belonged to. The Pz.Kpfw.IV on the right, tactical number '26' is missing two return rollers and one of the armoured shutters on the commander's cupola. The Balkenkreuz and tactical number are painted in white, the rhomboid below the tactical number (and below the Balkenkreuz on the rear of the Pz.Kpfw.III turret) are painted in yellow.

L.Archer

Right: An unusual view of a Pz.Kpfw.IV Ausf.C destroyed in Caribrod, today Dimitrovgrad, Serbia. An explosion has blown away the side of the superstructure and buckled the top plate, the remains of the track guard snake along the side of the tank. What at first looks like a bottle on the engine deck is probably a mount for carrying a spare roadwheel. Note the smoke grenade discharger fitted to the exhaust muffler and the opened port (for disengaging the fan drives while starting the engine) on the rear plate.

L.Archer

In front of a pair of Pz.Kpfw.I Ausf.A on a prewar training exercise in 1938 or 1939, this Pz.Kpfw.IV Ausf.A has features that were designed for the 14.5mm thick armour like the 'bent' front hull plate, machine gun ball mount, and driver's visors.

W.Auerbach